Health and Safety
MINES AND QUARRIES
1977

South Midlands District

LONDON: HER MAJESTY'S STATIONERY OFFICE

ISBN 0 11 882020 6

Contents

Foreword

The application of the Health and Safety at Work etc Act 1974 has widened the scope of the Mines and Quarries Inspectorate. Notable amongst the new entrants to legislation are landfill sites on old mine or quarry premises and exploratory drilling sites. A fatal accident at one of the latter sites indicated the wisdom of bringing allied activities into the field of the specialist inspectors who can apply the relevant legislation together with the appropriate accident prevention measures.

It was disappointing to find an increase in the number of accidents during the year, particularly those due to falls of ground in mines. This subject has been extensively studied in the past, the causes and effects are well known, and as no new hazard was revealed it appears the old lessons have been forgotten or not passed on to the new generation of miners. Also, despite considerable publicity and safety propaganda, there was an increased number of accidents due to haulage and transport activities.In coal mining the average age of underground workers has been gradually rising over a period of some 20 years, but more recently this trend has been reversed by a rapid intake of young men and the early retirement of men with long service. Encouraging as this may be in some respect, we must not forget that the early retirement scheme is taking out of the industry many man-years of valuable mining experience.

While they wait their turn for coal face training, new miners are usually employed on haulage and transport work in small teams of four or five men. If each team includes sufficient experienced workers to supervise and instruct the new men all is well, but investigation has found this to be the exception rather than the rule. The reason is easy to see. The influx of inexperienced miners now accounts for one-sixth of the total labour force and nearly all of these men are working underground on haulage and transport duties. If the person directing the deployment of men is not careful, a transport team may be comprised entirely of men with inadequate experience or training and the situation brought to light literally by accident. It appears this whole aspect of training needs revision and bringing up to date.

Looking on the brighter side, mining technology continues to improve as new machines and techniques are introduced. This is particularly true in the field of electronics which extends from the radio control of coal face machinery to closed circuit television in some mines and quarries, together with computerised control of many operations. Remote sampling and the recording of environmental changes can give early warning of danger and portable instruments to detect gases other than methane are being developed. In quarries much use is made of portable radio communication equipment and electronic exploders are now available for special applications where intricate blasting techniques are necessary to avoid damage or public complaint.

The technological improvements must be accompanied by better training, supervision and discipline if the worsening trend of the accident statistics is to be reversed in the coming year.

F. TOOTLE

HM Senior District Inspector of Mines and Quarries

1 General

1   The South Midlands District includes mines and quarries in the counties of Leicester, Northampton, and Warwick, and part of the counties of Nottingham, Derbyshire, West Midlands and Cambridge. The District boundaries were not changed during the year.

2   At the end of the year 29 mines and 246 quarries were working. Of these 27 mines were producing coal and operated by the National Coal Board and two were producing gypsum. The last of the fireclay mines ceased production during the year and was closed. The quarries include 11 opencast coal sites as detailed in Part 5 of this report.

3   During the year four persons were killed and 67 seriously injured in mines and quarries compared with three and 67 respectively in 1976. The total figures mask an increase of 70% in the falls of ground category and 30% increase in the haulage and transport accidents. Detailed statistics of all accidents are given in Appendix 2 and in the relevant sections of the report.

4   Staff in post at the end of the year is shown in Appendix 1. In February Mr. K.L. Twist was appointed to the post of HM Inspector of Mines and Quarries to fill a vacancy in the South Nottinghamshire part of the District, but in December Mr. B. Langdon, HM Inspector of Mines and Quarries was transferred to the Scottish District.

5   All mines and quarries were inspected at least once during the year. Each accident and dangerous occurrence reported under Sections 116 and 117 of the Mines and Quarries Act 1954 was investigated and appropriate action taken to avoid recurrence. Six inquests were attended and ten complaints relating to the working of mines and quarries were investigated. Of the complaints, seven were considered to be justified and remedial action was taken where this fell within the jurisdiction of the Health and Safety Executive, or the information was passed to the appropriate authority.

6   Details of inspections made by HM Inspectors are shown below. In addition, 129 inspections were made at tips, 78 at landfill sites, 54 under the Offices, Shops and Railway Premises Act 1963, and 158 at other premises including exploratory drilling sites.

Table a.   Inspections made by HM Inspectors of Mines and Quarries

	Coal mines	Other mines	Quarries	Total
Underground	835	14		
Surface	174	3	461	
Total	1 009	17	461	1 487

Source: Health and Safety Executive

7   HM Inspectors continued to take an active part in meetings of local branches of their professional institutions, and in meetings, safety competitions, first aid and firefighting competitions throughout the year.

8   The numbers of inspections made on behalf of workmen are shown below. Again this year, while several inspections were carried out at each mine at only five of the 246 quarries did the workmen and their trade unions take advantage of their right to make safety inspections. Safety is not the prerogative of the employer, everyone in a mine or quarry is involved and in particular, an employee has a vested interest in obtaining and maintaining safe working conditions.

Table b.   The Mines and Quarries Act 1954, S123   Inspections on behalf of workmen

Coal mines	Other mines	Quarries
723	6	12

Source: Health and Safety Executive

9   No prosecutions were instituted by the HSE during the year but disciplinary action was taken by the manager of one safety lamp coal mine when an underground worker was found with smoking materials in his possession in the mine. With the increased number of inexperienced men of all ages now entering the industry, it is worthwhile to point out that the danger of someone attempting to smoke underground is still present and the statutory searching procedures must be diligently carried out at all times.

2   Coal Mines

10   The number of mines producing coal remained at 27, all operated by the NCB. While saleable output fell from 17 300 000 to 16 600 000 tonnes, the number of men employed increased from 28 600 to 29 200 by the end of the year. The reduced productivity was the result of many factors. A successful recruitment drive to replace men leaving the industry under the NCB's early retirement scheme brought many men back into mining in addition to new entrants, but it takes a long time to make men into competent miners. That was not the only problem at two large mines where geological troubles and spontaneous combustion restricted production for half the year. Other mines also had labour troubles during the latter part of the year when the introduction of local productivity bonus schemes were being debated.

11   It is typical of the extractive industries that the best and most accessible mineral is worked first leaving the more difficult and less economic mineral behind. Several mines in the District are now working these poorer seams

although substantial reserves have been proved beyond the existing mining areas and development is eagerly awaited.

Even the comparatively new mines are becoming extensive with long haul roads which have to be manned and maintained in good repair, adding to both the productivity problem and accident potential simply by increasing the number of men at work on roads.

Accidents and their prevention

General

12   Four persons were killed and 63 seriously injured compared with three killed and 57 seriously injured in 1976. An increase of 14 underground accidents, equally divided between falls of ground and transport categories outweighed, in the general total, a reduction of seven accidents in the miscellaneous group. Surface accidents were reduced by one.

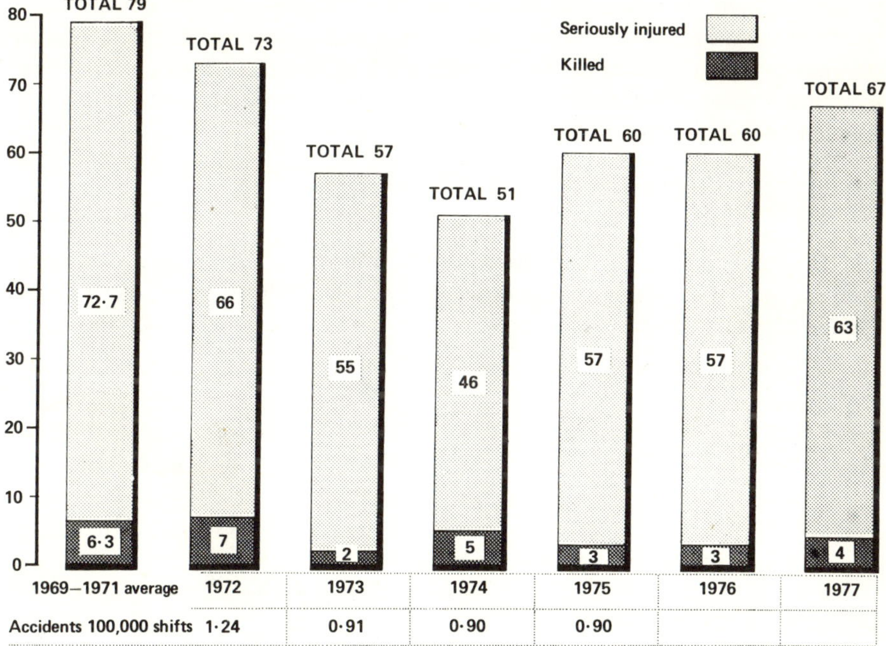

Fig 1   Coal Mines: Number of fatal and serious (reportable) accidents.

13   The longer term position is shown in Fig 1 which indicates, in recent years, a disquieting situation and reinforces the view that even more effort must be well directed if recognisable improvement is to be achieved. Three of

the four fatal accidents could be linked to the fact that moving equipment was either not stopped or not properly isolated after being stopped. This hazard was the subject of a written personal safety message from employers to individual employees last year and is so important that a concerted effort to press the approach even further would be well worthwhile. Human behaviour figures so prominently in many of the accidents that renewed attempts to communicate safety messages, over a wider field, should be made. The most effective, up to date, means of doing so should be investigated and exploited including audio and video tape techniques. A pilot scheme will be assessed during 1978. This approach is additional to the more technical preventive measures which are afforded considerable space in last year's report. It should not be underestimated on the ground that it has been tried before. There were few novel causes of accidents but this does not mean that prevention techniques, whether human or technical, should not be updated with an enthusiasm that will bring its own reward.

14   It is also clear that a considerable proportion of accidents is concerned with equipment and machinery that, in the broader sense, fall within the province of mechanical engineering. I recommend that the persons of this discipline should be used to complement existing safety staffs. Their specialised knowledge and training could make a valuable contribution to accident investigation and prevention.

Accidents from falls of ground
(25·4% of all persons killed and seriously injured)

15   One person was killed and 16 seriously injured during the year compared with one killed and nine seriously injured in 1976. This reverse, both in the immediate and longer term record, is illustrated in Fig 2. A more

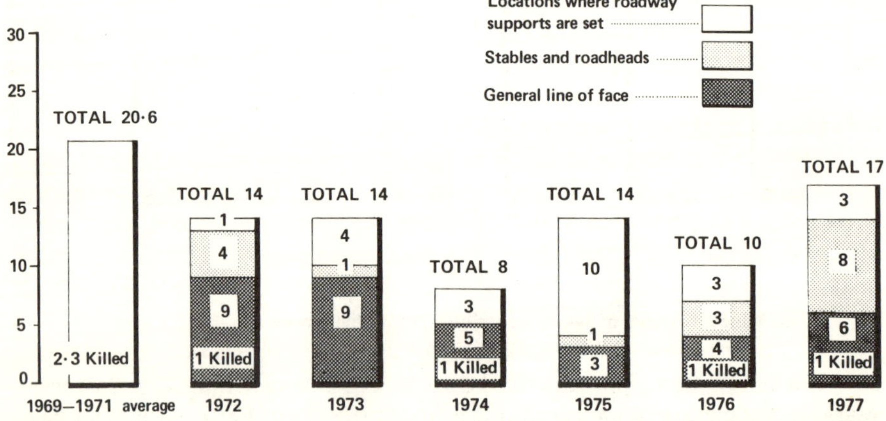

Fig 2   Coal Mines: Accidents due to falls of ground.

comprehensive analysis of the immediate position is shown in Table 3, and indicates that the increase in number of accidents from last year was in locations where face supports were set. More particularly the increase was greater in the face end areas of power loaded faces, especially at loader gates, and mars the good record established in recent years for this location. Although no facet of support can be neglected, it remains clear that the main hazard exists in the areas where newly exposed strata is not supported soon enough or well enough. Until positive improvements are secured in this area of work, accidents will occur whether the main support system utilises powered supports, props and bars or steel arches.

Table c. The location of accidents from falls of ground

		Falls		Total	
	Location	*Face/ side*	*Roof*	*1977*	*1976*
Face supports in use	(a) The general line of power loaded faces excluding (b)	3	3	6	4
	(b) The face end areas of power loaded faces excluding (c)	3	5	8	3
Roadway supports in use	(c) From ripping lip of power loaded faces to 10 m. outbye	—	—	—	—
	(d) Advanced headings	—	—	—	—
	(e) Development headings	1	2	3	2
	(f) Elsewhere underground	—	—	—	1
All locations totals		7	10	17	10

Source: Health and Safety Executive

Locations where face supports were used

16   Six persons were seriously injured on the general line of power loaded faces compared with four in 1976. Four of the accidents occurred in the prop free front area where men are still necessarily exposed to hazard despite the technical progress made towards a man free front. Two occurred in the travelling way through supports. When men are required to work in the prop free front area protection must be provided. Arising from the most serious of these accidents it is imperative that managers stipulate and enforce a systematic support system where maintenance or repair work is carried out on machinery in the prop free front area. Such a scheme would be of optimum practical value if it was based on built-in powered support rather than on hand set props and bars. Two accidents involving broken arms were sustained by a shearer operator and cable handler respectively by falls of face in a working more than 2m high. Such accidents could be avoided or minimised if more effective deflector guards were fitted to machines and if more use was made of purpose designed reinforced gauntlets.

17   At the face end areas of power loaded faces, one man was killed and seven seriously injured compared with three persons seriously injured in 1976. Seven of the accidents occurred at loader gates. The fatal accident occurred where a gathering arm loader was installed and resulted, as did one of the serious injuries, from lack of temporary support beyond the last permanent support. Three serious injuries were associated with stable elimination systems to men exposed in an inadequately supported prop free front area for purposes of maintenance, cleaning up of remnant coal and bar setting respectively. If stable elimination is to continue to make a contribution to accident prevention, as it is extended through a wider range of conditions, every effort must be made to maintain the best standards available with powered support of newly exposed roof and, wherever practicable, face. The remaining two accidents resulted from falls of roof sustained by men passing through the underip area, and might have been prevented by the extended use of powered supports or lagging.

Locations where roadway supports were used

18   Three accidents occurred at the face of development headings compared to two in 1976. Two occurred in fully mechanised headings involving respectively falls of roof and side in advance of the last permanent arch and might well have been avoided if the best standards of support protection known in the district had been applied. The third accident occurred to a workman who was standing in the bucket of a loading machine coupling an arch when he was struck by a fall of face. The practice of standing directly in a loader bucket should be prohibited and alternative safe staging substituted as a well known accident prevention measure. Progress on this specific point has been initiated and experience will be evaluated in the coming year.

Underground transport accidents
(41·8% of all persons killed and seriously injured)

19   One person was killed and 27 seriously injured compared with 21 persons seriously injured in 1976. This deteriorating situation, illustrated in Fig. 3, in a category which has been highlighted as being the most prolific accident agency, must serve to confirm the pressing need for conscious and determined effort to change matters. Managers, officials and workmen should accept, without reservation, that almost without exception, such accidents are avoidable and be motivated by this realisation. The circumstances surrounding many of the accidents suggest very strongly that human failings played a prominent part that cannot be ,overlooked. These failings range from inadequate specific training through apparent oblivion to danger to blatant unsafe behaviour that must be the responsibility not only of the persons concerned but also of those who condone it until an accident occurs. From a

technical aspect it would appear that many haulage engines are used with rope pull capabilities well in excess of normal requirements. Load limitation on haulages used for installation and salvage is generally accepted and as an interim protection for supplies haulages I recommend the development and installation of excessive rope pull warning devices. Further, the popular types of supplies haulages should be assessed with regard to the degree of control that is available to the driver particularly on starting.

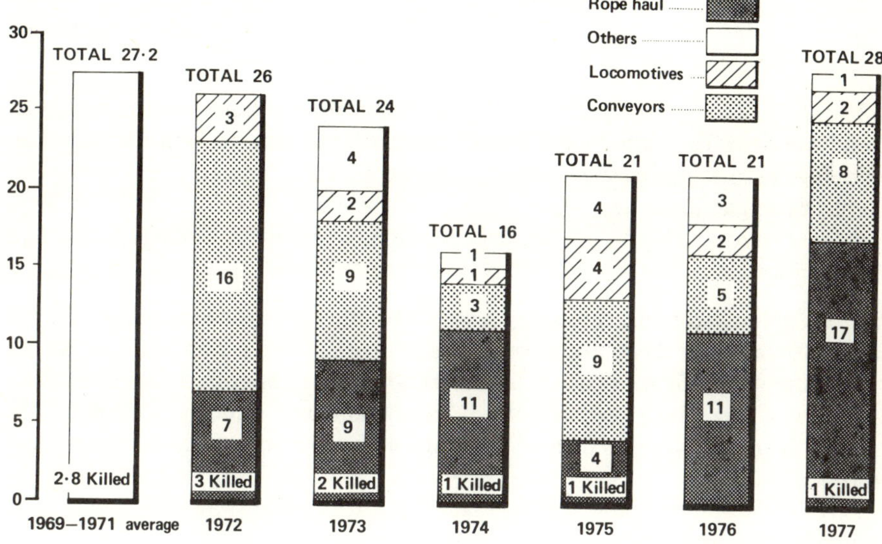

Fig 3 Accidents due to haulage and transport.

Accidents on rope haulage systems in roadways

20 Seventeen persons were seriously injured by rope haulage systems in roadways. Of these, thirteen involved wheeled vehicles running on rails and four involved the hauling of equipment by rope haulage without using wheeled vehicles. Of the latter, three accidents resulted from the use of relatively powerful haulages to haul equipment including a loader bucket and a shearer drum, without the use of vehicles or track. Such practice should be prohibited in transport rules, and where the provision of a suitable track is impracticable, then a more controllable and less powerful haulage medium should be used.

21 Four accidents involving vehicles resulted from violent movement of ropes mainly due to inadequate rope control arrangements at places of localised change of gradient and direction coupled with a failure to reduce the effects of such changes by a process of regrading. A particularly serious accident in this category occurred when an over loaded set of vehicles was

being hauled up a drift rising at 1 in 3. As a result of additional shock loading, arising from inexpert operation of the haulage engine, the hook type rope attachment straightened, became detached and sprang violently causing severe head injuries to an accompanying workman. Every effort should be made to exclude open hook rope attachments from haulage systems.

22   Three persons were seriously injured by derailed vehicles. The inadequate standards of track laying sometimes found during inspection were not a main feature of the accidents which were caused more immediately by adverse operating characteristics of the haulage engine, compounded by workmen positioning themselves in potentially dangerous locations.

23   Of the five accidents caused by contact with moving vehicles or loads, four must be realistically attributed primarily to unsafe attitudes of men ranging from sheer carelessness to undoubted indiscipline. The exceptional accident might have been prevented if the practice of siting signalling keys adjacent to refuge holes, particularly at places where men regularly work, had been adopted.

24   A serious accident caused by a set of 14 mine car type vehicles which ran away for a distance of 140m on a gradient of 1 in 28 was initiated by an unsecured load of rails which fell onto and loosened the rope attachment. The location was a place regularly used for marshalling vehicles and such sites should be regularly inspected with emphasis on protection against forseeable hazards.

Conveyor accidents

25   One person was killed and seven seriously injured by conveyor accidents. Three of the accidents were associated with chain stage loaders, two with armoured face chain conveyors and one with a roadway chain conveyor. Two persons were seriously injured in accidents associated with belt conveyors.

26   The fatal accident and one serious accident caused by stage loaders were associated with taking supplies from the gate to the face. The former involved manhandling long roof bars under the ripping lip and the latter the reversal of the stage loader to convey wooden blocks. Transport rules should stipulate that conveyors be stopped and locked out when awkward or heavy equipment is being manipulated in the vicinity and the reversal of the chain conveyors for supplies transport should be similarly prohibited.

27   In a serious accident involving a moving armoured face conveyor, a fitter working on a shearer lost a leg when a prop which had accidently dropped on to the conveyor became caught in the shearer underframe and moved the shearer violently into the ribside. When persons have to work in a location of such high accident potential, the other working machinery should be isolated.

This was the subject matter of the personal safety message already referred to which managers should enforce.

28 A serious belt conveyor accident occurred to a 17 years old patrolman who was found injured close to a transfer point, after a blockage had been caused by a steel girder being transported outbye. The circumstances must make clear to all concerned with the training that the statutory 20 days close personal supervision is a minimum that may well be inadequate for particular persons to be left to do particular jobs in particular situations. Close personal supervision should be followed up by close personal interest particularly for younger employees.

Locomotive accidents

29 Of the two accidents in this category, one occurred during legal manriding when a derailed vehicle overturned on a worn crossing. Two previous derailments on the same shift, at the same place, should have lead to the cessation of manriding until repairs had been completed. The other accident occurred when a passing captive rail locomotive caught and moved a loose conveyor pan which trapped a deputy thus demonstrating that new systems cannot make up for human shortcomings.

Accidents from machines
(10·4% of all persons killed and seriously injured)

30 One person was killed and six seriously injured in accidents involving machines, continuing the adverse trend shown in Fig 4.

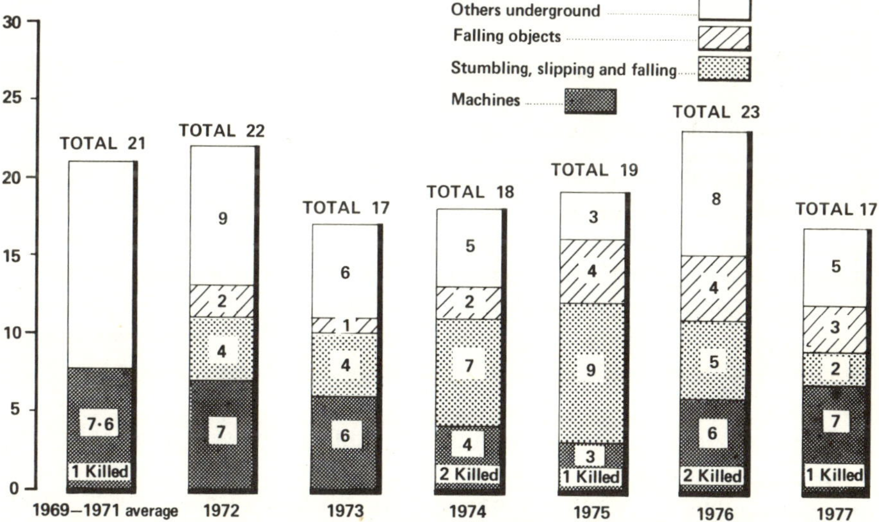

Fig 4 Accidents due to machines and miscellaneous causes.

31   An 18 years old haulage worker was killed at a shaft bottom tippler site when he was crushed between the roadside and the full mine car, which only partly entered the tippler, when the tippler unexpectedly started to tip. There was failure to carry out correct isolation procedure before attempting to move the car over the rail level mounted wheel operated starting mechanism.

32   Two serious accidents occurred during the operation of heading loading machines. One involved a continuous miner cutting boom which slewed when a pin sheared in one of the slewing chain bollards and the other occurred when a gathering arm loader ran over the operator's leg. The best designs of modern loading machines site the operator within the confines of the machine where risk of such accidents is less. A proved modification for the gathering arm loader was illustrated in the Annual Report for 1972 and should be more generally adopted. There will always remain the need for persons to exercise prudence when positioning themselves close to such machines.

33   Three accidents occurred to men struck by oscillating power loader haulage chains. The introduction of chainless haulage systems is proceeding but in the meantime continued attention should be paid to the use of chain restrainers and to the maintenance of good face alignment on faces where chains are installed.

Accidents from miscellaneous causes
(14·9% of all persons killed and seriously injured)

34   Ten persons were seriously injured compared with 17 seriously injured in 1976. The trends in the main categories of miscellaneous accidents are also shown in Fig 4.

Stumbling, falling and slipping

35   Two persons were seriously injured in this category which shows a progressive and welcome improvement over the last three years. A staging being used as a platform for the erection of supports collapsed to cause one accident and the rung of a wooden ladder snapped to cause the other. Wherever practicable steel ladders should be used for access.

Falling objects

36   Three persons were seriously injured by falling objects. Two might easily have been avoided if more care had been taken in dismantling and loading salvaged equipment. Another, more unusual, resulted when a warwick girder fell from its bracket which had suffered previous damage.

Other accidents

37   Two persons were seriously injured in accidents when stage loader advancing rams were being operated. A sound design of these essential pieces of equipment has now been specified by mechanical engineers in the South Midlands Area of the NCB and should be utilised generally.

Surface accidents
(7·5% of all persons killed or seriously injured)

38   One fatal and four serious (reportable) accidents occurred on the surface. The generally satisfactory longer term trend, shown in Fig 5, was marred by a fatal accident to a 63 years old man in a coal preparation plant who was killed when a stationary conveyor on which he was working started up. This and another accident on a shaft top traverser might have been avoided by proper isolation procedure.

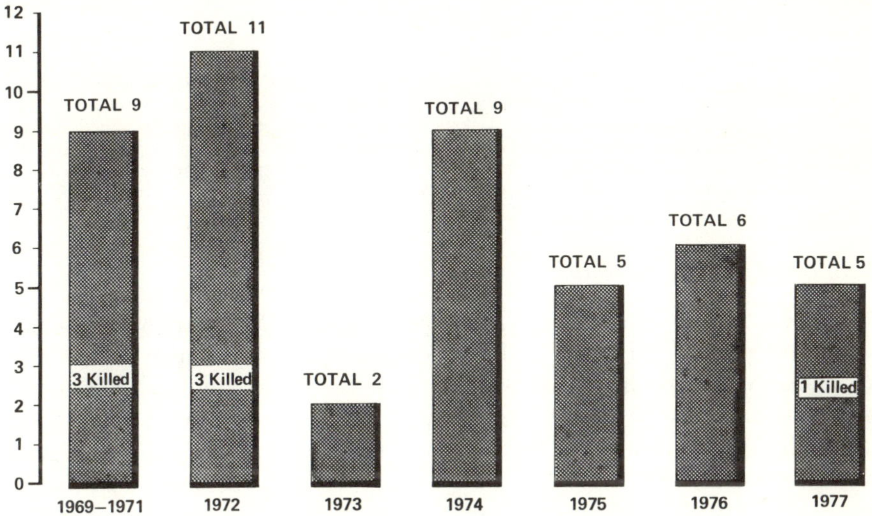

Fig 5   Accidents on the surface, coal mines.

39   When the side of a scrapped railway wagon was sufficiently weakened by burning and existing corrosion it collapsed in a gusty wind causing serious injury to a fitter standing inside the wagon. This not unusual task on a mine surface is carried on in more remote surroundings and is the type of operation to which a succinct set of rules could be applied.

Dangerous occurrences

40   There were seven dangerous occurrences reported during the year compared with 10 in 1976 and an average of 11 in the previous three years. It is

pleasing to note how the number of dangerous occurrences has declined and much of the credit must go to the management and engineering staff at the mines. All the occurrences were fully investigated and were classified as listed below:

Table d. Dangerous occurrences in coal mines

Type of occurrence	1976	1977
Ignitions of gas or dust below ground	2	1
Outbreaks of fire below ground	—	2
Withdrawals of men owing to smoke, etc.	1	1
Breakdown of ventilating apparatus	—	1
Overturning or failure of lifting machines	1	—
Electric shock or burns	4	1
Injuries from blasting materials or devices	—	1
Unstable or potentially unstable waste heaps or settling ponds	1	—
Use of apparatus in pursuance of the Mines (Emergency Egress) Regulations 1973	1	—
Total	10	7

Source: Health and Safety Executive

NOTE: During 1976 and 1977 no dangerous occurrence was caused by accidental ignition of gas on the surface, fire on the surface, outburst, inrush of gas from old workings, inrush of water, breakage of man-carrying ropes, etc in shafts, staple pits or unwalkable outlets, breakage of man-carrying ropes below ground, overwinds, explosions from, or collapse or burst of pressure vessels, collapse of certain surface buildings or structures, first aid or medical treatment arising from use of breathing apparatus etc, or failure of breathing apparatus.

Ignition of gas below ground

41   Firedamp was ignited in an advanced heading shortly after a round of shots had been fired using a P4/P5 type explosive but the flame self-extinguished almost immediately and no person was injured. Subsequent investigation revealed minor malpractice in the shotfiring procedure and incomplete detonation of the explosives in certain circumstances. A P5 type explosive is now being used for shots in coal. At many mines a P4/P5 type explosive is used for the dual purpose of firing shots in coal and delayed action shots in rippings in order to simplify the system of storage, distribution and use of explosives and detonators. The simpler the system the better as mistakes are less likely to occur, but as in all cases of compromise there are both gains and losses overall and in shotfiring, simplicity cannot be achieved at the expense of efficiency.

Outbreaks of fire below ground

42   Two outbreaks of fire were reported during the year. In one case an electrical switch panel controlling a gate belt conveyor burst into flames following several short periods of reverse operation of the conveyor. It was

apparent that a short circuit had occurred in the switch and that the fault had not tripped the power transformer protection circuits, but the equipment was too badly damaged to determine the cause of the original fault. As conveyors or other machines may be repeatedly started in either direction, the switchgear must be made and maintained to accommodate repeated operation or designed to enforce a time lag between changes of direction. The second outbreak of fire occurred at an underground coal bunker station where it is suspected sparks from a defective motor coupling in the hydraulic power pack ignited coal dust and oil from the emulsion used as the fire resistant fluid in the hydraulic circuit. This case is dealt with in more detail in the mechanical engineering section of the report.

Withdrawal of men owing to smoke etc.

43   Men working on a longwall face had to withdraw, wearing their self-rescuers when smoke and fumes contaminated the ventilation. On investigation the cause was found to be at a gate belt conveyor drivehead where a thruster brake had failed to lift off and a motor bearing had collapsed. On further investigation it was found the brake was not necessary and it was completely removed. Almost every year at least one similar incident occurs in the District but the lesson is not learned. Some thruster brakes are installed simply because they are purchased as part of the drivehead package. If the brake is necessary it must be properly installed, protected by circuits which do not allow the conveyor to run with the brake on, and must be maintained and adjusted to a high degree of accuracy. If the brake is not required it should not be installed, or if circumstances change it should be taken away as soon as it becomes unnecessary.

Breakdown of ventilating apparatus

44   During a holiday period the main fan had been stopped for maintenance when the standby fan broke down, leaving the mine without positive ventilation. When a machine as vital as a fan is only used occasionally it must be tested and run regularly to ensure it is in good working order and in fact available for use.

Electric shock or burns

45   After a spate of incidents in 1976 only one was reported during the year. An electrician working in a pit bottom substation burned his hands when he was recoupling a cable and a flash-over occurred. The investigation revealed that the electrician in charge had operated an earthing lever to make doubly sure the cable was dead but due to a malfunction of the switchgear this action had in fact restored power although the main switch remained in the 'OFF'

position. The incident shows that when working on high tension circuits nothing can be left to chance and strict discipline is necessary to ensure a circuit is actually isolated and locked off before any work is allowed to commence.

Injuries from blasting materials or devices

46   An apprentice electrician was slightly injured by material projected from a shot fired some 44m away in a roadhead. Shotfiring curtains can do much to avoid this type of accident but it is too difficult to prevent all small fragments from passing through. Men cannot rely on the curtain alone for protection and it is the shotfirer's duty to see everyone in the danger zone has taken shelter. This is particularly true when apprentices and other young persons are in the vicinity.

Mechanical engineering

47   The implementation of the recommendations of the National Committee for Safety of Manriding in Shafts and Unwalkable Outlets was continued during the year, and detailed programme with priority action defined have been planned to cover the next five years.
Steady progress has been made out but obviously any new installation must comply with the recommendations. When two steam winders and an old electric winder were replaced by electric engines transferred from shafts or mines which have been closed, the opportunity was taken to refurbish the braking system and eliminate single line components.

48   Even when single lines in winding engines brake systems have been eliminated it is important that in the event of failure of any component at least half the braking effect remains available. In most cases this means that equal braking effort must be produced on each side of the drum, but tests have shown that this is not always so and one side provides more than half the braking force. If a failure occurred on this side then the remaining brake force could be insufficient to control the landing speed under emergency conditions. To avoid this danger, some winding engines brakes have been fitted with load sensing devices which monitor the forces in each half of the braking system, enabling accurate adjustment and indicating the state of balance of the brakes.

49   Two further installations of pit bottom buffers were completed and more will be included in pit bottom reconstruction work.

50   At two mines it has been demonstrated that winding mineral in mine cars without using keps is practical and can result in a reduction in decking times and there is no technical reason for retaining keps for manriding. Many shafts are now equipped with interlocking arrangements required under terms of

exemption permitting winding without keps but generally, progress has been slow in this direction.

51   No dangerous occurrence involving shaft winding was reported during the year and there was a reduction in the number of serious though non-reportable incidents. This is a very welcome improvement although there are no grounds for complacency as the following incidents show. When an ascending cage carrying six men failed to enter the surface receiving structure after an overspeed trip, it was found that a hinged platform carried on top of the cage had become dislodged. The platform which was used for shaft maintenance was removed and now is only fitted when required.

52   At a small upcast shaft where occasional contact with the shaft wall had been reported, cage clearances have been made adequate by fitting a smaller cage, re-tensioning the guide ropes and fitting an additional guide to reduce the cage movement during winding.

Haulage and transport

53   During the year there has been considerable increase in the number of rope hauled manrider installations including a trapped rail type haulage at one mine where roadway floor conditions were difficult. The installation of additional manriding facilities inevitably increases the work load on maintenance craftsmen as the high standards demanded at the time of installation must be maintained and sufficient competent staff must be employed to carry out inspections and repair work.

54   When a scoop coupling is used for speed control of a manriding haulage engine the motor itself cannot provide any braking effect neither mechanically nor electrically and this can be a problem on steep gradients. An unusual method of providing speed control and braking assistance is being tried at one mine where a manriding haulage engine is being driven through a gear box which though constantly engaged, allows any one of five reduction gear ratios to be selected by hydraulically operated clutches. The engine is also fitted with a torque limitation device and a comprehensive monitoring system to detect any malfunction of the gear box, clutches or the torque limitation device. If the trials are successful and the drive is proved to be reliable it is intended to fit radio control equipment and operate the engine from within the manriding train.

55   At two mines small battery powered locomotives with rubber tyred wheels have been introduced. These locomotives are not intended for main haulage work but have proved to be extremely versatile where conditions permit and can replace an assortment of rope haulage systems. One locomotive was used to haul powered roof supports from the shaft sidings into position on the coal

face, eliminating many of the hazards associated with transport and installation operations.

Fire resistant fluids

56   An outbreak of fire, previously referred to in the dangerous occurrences section, occurred at an underground bunker power pack. The investigation revealed that the hydraulic fluid emulsion had deteriorated in service and a subsequent general survey showed that in some installations the water content of the fluid had fallen from the original 40% to as low as 15%. When any hydraulic system is being filled or topped up care must be taken to ensure the correct fluid is used and in the case of water/oil emulsions, samples should be taken regularly to check the water content. The incident also shows that water/oil emulsion spillage rapidly dries out leaving an oil rich residue and if the power packs are not kept clean this residue mixed with coal dust produces an excellent fuel in the event of an outbreak of fire.

57   In recent years the highly flammable mineral oil in underground hydraulic machinery has been largely replaced by fire resistant fluids, particularly at shaft sides and in main intake roadways where the fire risk is obvious. However, in an attempt to reduce the inefficient use of compressed air, particularly associated with cage decking equipment, some systems are being replaced by hydraulic devices operating on mineral oil. This retrograde step of introducing new hydraulic equipment which has not been designed to operate on fire resistant fluids is a matter of some concern, and every effort must be made to convert or replace this machinery.

General

58   A central underground hydraulic power station to provide hydraulic power at several faces was installed in one mine during the year. An unusual feature of the installation is the use of reciprocating pumps having separate outlets within the same units which can supply fluid at different pressures according to requirements whilst still maintaining balanced forces at the crankshaft. There appeared to be many advantages in this arrangement, namely, no pumps or tanks to be advanced in the gate roads; fewer prime movers involved; better control and supervision of the equipment and fluid quality; less possibility of unauthorised alteration of the pressure regulating valves associated with the face hydraulic supplies, and standbye capacity is available without the need to duplicate equipment at each face.

59   The Chief Inspector's Report for 1976 referred to the fitting of a water cooled exhaust pipe to a diesel locomotive. A similar device was successfully fitted to a locomotive at one mine in the district and demonstrated the effective

reduction of the surface temperature of the hot exhaust pipe. The results indicated one practical means of modification of older types of diesel locomotives which do not comply with the current requirements regarding limitation of surface temperatures.

60   A surface reciprocating air compressor was severely damaged when part of a discharge valve assembly fell into the upper part of the vertical cylinder. Investigation revealed that the retaining device used to hold the valve in position had not been properly secured, had worked loose and subsequent vibration caused the valve assembly to break up and fall into the cylinder. The incident happened only 24 hours after completion of routine maintenance during which the valves had been dismantled and examined but there was no detailed record of the maintenance work carried out. It was found that the appropriate check list for reciprocating compressors was not available at this or other mines in the Area, nor had the compressor been examined during the course of each shift, as later arranged, or the defect would have been noticed before the machine was wrecked.

61   An accident at a surface coal preparation plant emphasised the danger of mixing metric and imperial screw threads, particularly when using lifting tackle. While using certified and apparently suitable lifting equipment to dismantle a pump a mechanic was injured when a section of the casing fell on to him and it was found that an eye-bolt had been pulled out of its tapped hole. The eye-bolt was of 20mm metric thread but the hole thread was seven-eighths of an inch Whitworth. Due to dirt inside the hole the eye-bolt could be tightened initially but when the hole was cleaned out and re-tapped it could not be retained. Whilst British Standards require eye-bolts to be stamped showing their thread form and safe working load no such requirement exists for tapped holes provided to receive eye-bolts. As many such holes were tapped sometime before metrication was generally practised, each presents a danger. It is recommended that all holes in castings and elsewhere which are tapped with a metric thread should be so identified on the casting. All personnel could then be instructed that any unmarked hole must be suspected of having imperial threads and carefully checked before any eye-bolt is inserted.

Electrical engineering

Winding

62   In the continuing process of modernisation, mainly the replacement of steam winding engines and renewal of old electrical winders, other improvements have been made to the ancilliary winding apparatus. Mercury arc rectifiers are being replaced by modern, solid state silicon controlled

rectifiers and much more attention is being paid to inter-locking, limit switches, testing and backing out controls and facilities for handling extra long materials.

63   There is renewed interest in automatic winding and automatic signalling in shafts, and trials are being arranged at two mines. Shaft communication systems have been fitted to the manriding cages at two other mines.

64   Winding without keps under terms of exemption is being encouraged, and in most cases the necessary inter-locking devices can be easily fitted. These inter-locking switches must be carefully sited however, as was shown by an incident in one shaft where a skip door automatically opened at the wrong time and caused considerable damage. Engineers must anticipate possible, even improbable. malfunction of equipment to ensure the system fails safe particularly in and around shafts. Where possible the safety circuits should not only give some indication that they are working but also an alarm signal when they are not. It may be too late to wait for an unusual reaction of the equipment to indicate a fault.

Haulage and transport

65   On conveyor coal clearance systems the use of computer controlled equipment has been extended to three more mines and some bunkers are now fitted with two speed out-loading conveyors which provide much improved control. Removal of the human element has generally improved conveying and bunkering performances in terms of flow and tonnage carried but the weakness of the system is still in the transducers on which the computer depends for the precise information necessary for its operation as programmed. Until the operation of the transducers is more reliable, conveyor attendants, patrolmen and engineers will have to be vigilant and check immediately any unusual operation of the equipment.

66   On rope hauled manriding haulages, the use of radio signalling apparatus is being extended and on two trains, alternators driven from the revolving axles of the carriage provide power for the head lights replacing the less effective portable battery lamps.

Coal face equipment

67   Various forms of chainless haulage systems are gradually replacing power loader haulage chains but almost invariably a cable handling problem is introduced as the traction devices and the cable are competing for the same limited space. Face cable handling cannot be ignored, it is part of the system and must be part of the original design.

Plate 1a   Hydraulically powered face sprag (Daw Mill Mine). (Reproduced by permission of NCB)

Plate 1b Hydraulically powered face sprag (Bagworth Mine). (Reproduced by permission of NCB)

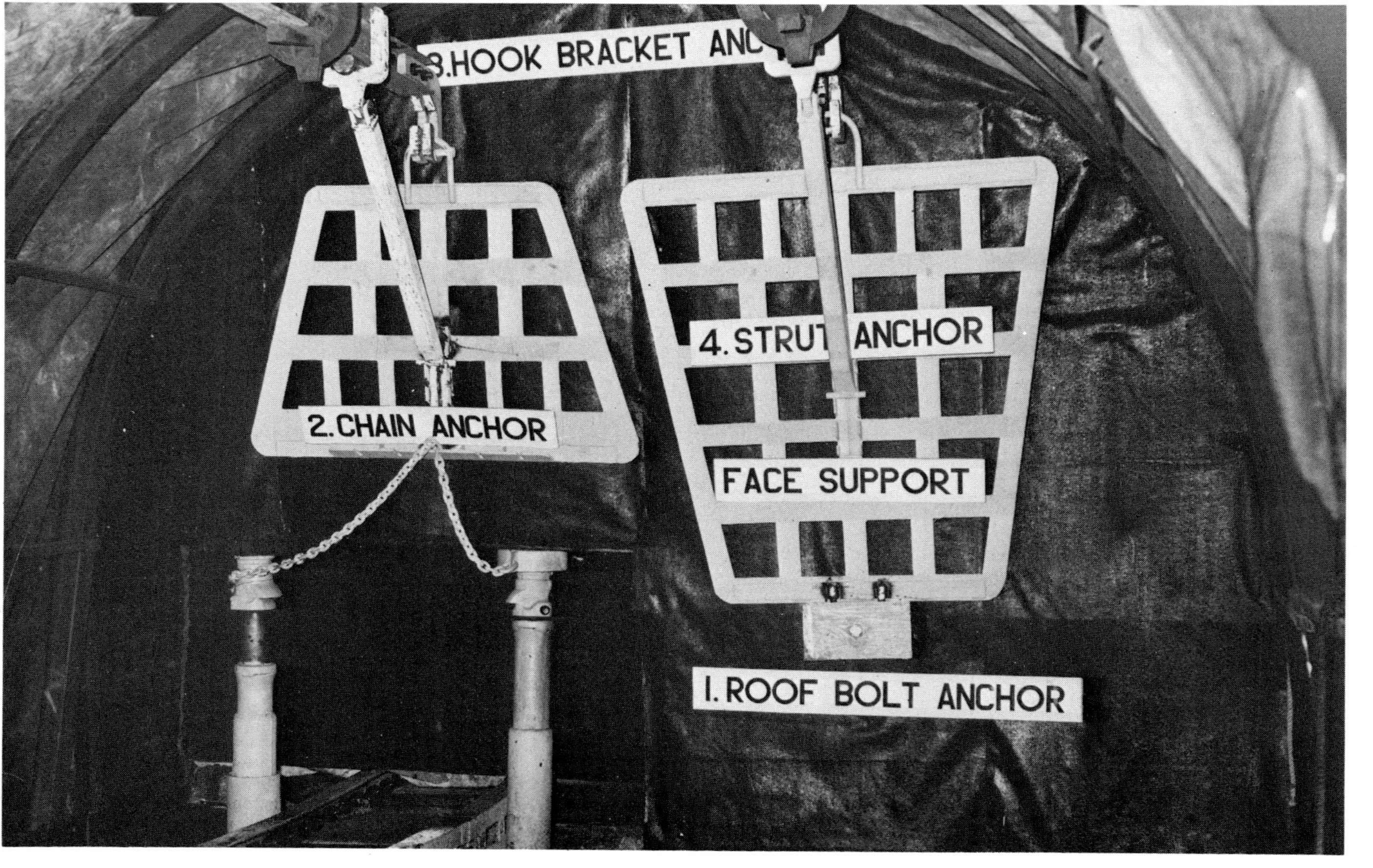

Plate 2 Hand set support for face of rippings or headings. (Reproduced by permission of the NCB)

Plate 3 Prototype curved creeper (Bentinck Mine). (Reproduced by permission of NCB)

Plate 4   Friction retarder for mine tubs (Whitwick Mine). (Reproduced by permission of NCB)

68   A proto-type system eliminating trailing cables to power loading machines is now being installed underground following extensive surface trials. One reason for the delay is that the trials proved a weakness in one of the rubber seals and modifications had to be made.

69   Face lighting has not made much progress this year mainly because of difficulties with connecting cables between the lighting units which failed to survive normal mining conditions. Some work has been done at one mine in an attempt to make the fittings more robust without affecting their intrinsic safety and results so far look promising.

70   At one mine a power loading machine has been fitted with radio control equipment but the working of this device has been erratic. Various modifications to make the equipment more reliable have been tried with limited success and so the work continues. This method of remote control was fully accepted by the men concerned who were quick to see the safety features of the system.

71   At one mine a new four unit gate end box switch was on trial and operating successfully. This was designed to reduce the mass of switchgear to be advanced with the face and at the same time give more reliable operation and better electrical protection.

Ventilation

72   Two cluster type booster fans have been installed underground during the year, and both were fitted with instrumentation for remote monitoring purposes. On a lesser scale where some auxiliary fans may run unattended, protective devices have been fitted to stop the fan in the event of excessive vibration, high bearing temperature, reduced air flow or more than $1 \cdot 25\%$ firedamp in the fan duct. The fan has to be restarted on site after inspection by an official, and not all such fans are so protected at present, but it is felt this is a move in the right direction.

73   As referred to in the Dangerous occurrence section, an electrical fault stopped a standby surface fan when the mine was dependent on it for ventilation. Standby equipment must be available and capable of replacing the equipment in normal use, not regarded as second best.

Power supplies

74   The demand for power at mines increases annually and it is almost a continuous operation renewing and up-dating power distribution equipment. At two mines new main sub-stations have been built and equipped during the

year, and new shaft feeder cables installed. The electrical linking of mines underground has also improved supplies in one case.

75   In the South Notts NCB Area it is now the policy to use as standard aluminium core, single wire armoured cable for all underground high tension supply lines. The cables have an outer PVC sheath to hold the single wire armour in place and they are much easier to handle. They have been in use for less than a year but so far their resistance to mechanical damage appears to be satisfactory.

General

76   Defective protective devices and internal short circuits resulting in damage to switch gear caused, in one case, an outbreak of fire below ground. Only the highest standards of installation and maintenance can prevent such incidents.

77   Various incidents throughout the year have indicated the need for improved training and frequent re-education of electricians and operators of electrical apparatus. On at least three occasions, wrongly fitted or defective components resulted in the dangerous condition that normal isolation procedure left the circuit live. This also emphasises the danger of relying on remote control switches or lock-out signals instead of the complete isolation procedure.

78   To give a final safety check after normal isolation procedures have been followed, dead line testers are now being issued at each mine in the District, but unless correctly used these will not prevent accidental contact with live conductors.

79   To reinforce the 'Permit to Work' system, personal locks and multi-locks have been introduced together with a written instruction regarding isolation of equipment under repair or considered to be in a dangerous state. The full application of the combined system should make it impossible to cause an accident by inadvertant operation of switch-gear and should mean a man, be he technician or labourer, may work in confidence that machinery cannot be started without his knowledge.

Ventilation and airborne dust

80   Although there was no major ventilation change in any mine during the year, the existing ventilation systems were improved at three mines by the installation of a new main fan at one, and underground booster fans at the other two.

81   Auxiliary fans have been under close scrutiny since the public enquiry into the causes of the Houghton Main Explosion in 1975, and standards of installation and maintenance are improving. The recently issued NCB directive has reinforced the requirements of the regulations and the Manager's Fan Rules should remove any doubt of interpretation from the minds of the men concerned. To this end, a standard form of rules has been evolved and is in use throughout the District.

82   All main surface fans and underground booster fans recently installed are covered by comprehensive monitoring systems which in effect provide continuous examination of the fan, its bearings and its drive mechanism and avoid the need to employ a fan attendant in a lonely and monotonous job.

83   At one mine workmen have been issued with automatic firedamp detectors instead of the usual flame safety lamps. There have been no problems and it is generally agreed to be more efficient and reliable system.

84   The re-circulation of air through dust filters is now standard practice where airborne dust problems exist, but it is again necessary to stress the need for caution. Where, subject to terms of exemption from the ventilation regulations, air is deliberately re-circulated monitoring devices will give warning of any build-up of firedamp, but if a free standing filter comprised of a portable fan with an inlet air dust filter is placed in a heading, the total ventilation of the heading may be adversely affected.

85   During the year 37 notifications of adverse airborne dust samples were received from mine managers in the District. Most were due to some temporary cause varying from a simple blockage of jets in the dust suppression equipment to major geological disturbances. On power loaded longwall faces in thin seams where horizon control is difficult, a slight deviation causes the machine to cut into the roof or floor with adverse effects on both airborne dust and the ash content of the coal product. Clearly it is in everyones's interest to keep the machine in the coal wherever possible, and better nucleonic steering devices are being developed to this end.

Fires and rescue

86   Fire prevention in mines may be divided into two parts. First the elimination of flammable materials as far as practicable and secondly, good standards of installation and maintenance of machinery and electrical equipment so that machinery will stand up to a certain amount of rough usuage without going on fire. Then if a fire does occur, the fuel in the vicinity is minimal. Certainly no flammable material should be taken underground if there is a non-flammable alternative, so it was very disappointing to find new hydraulic equipment installed and working underground with mineral oil as

the hydraulic fluid. More is said about this in the Engineering section of the report. Installation and maintenance standards vary from place to place but particular attention must be paid to known fire hazards such as conveyor brakes and portable hydraulic pumps.

87   Operators of machinery should also do their share of fire prevention work by keeping machinery reasonably clean and distinguishing between its use and abuse. Fire prevention is always better and less hazardous than fire fighting.

88   At all NCB mines in the District fire officers have been appointed to take charge of fire fighting arrangements, fire drills and the maintenance of fire fighting equipment. Generally, on inspections standards were found to be good and the equipment properly sited. On the surface, some progress has been made with the issue of fire certificates under the Fire Certificates (Special Premises) Regulations 1976, and meetings of fire officers have been held to discuss and implement this recent legislation.

89   The District continues to be served by Rescue Stations at Arley (South Warks.), Ashby de la Zouch and Ilkeston. Emergency winders are maintained and operated by the two latter stations. Permanent staff at the three stations include nine full-time officers, 21 permanent corps men and two instructors. Four mines in the South Notts Area use the training facilities at Mansfield Rescue Station.

90   Again, the recruitment of part-time brigade men exceeded retirement, and at the end of the year 311 trained men were employed at coal mines in the district with a further nine at Gypsum mines. Thirteen requests for assistance were responded to by the Rescue Stations. The incidents mainly involved spontaneous heating as several mines in the district are particularly prone to this hazard.

91   Self-rescuers were used on four occasions involving a total of thirty-four persons. The life saving potential of this apparatus is most appreciated by those who have needed to wear it, and it is of the utmost importance that the standard of training in its use be upheld. Replacement of the type 265 rescuers with the type 275, which has a longer protection period of 90 minutes, continues.

92   The development of the inflatable stopping has proceeded and in the coming year the working party will continue their appraisal of its value. Concurrently, they will carry out trials with a smaller inflatable stopping developed in the South Wales District and will design a model of intermediate size.

93   Roadway dust treatment by chemical consolidation and application of incombustible dust is generally effective in mines in the District, and good use is made of electrical and diesel powered stone dusting machines. There is no mistaking a roadway treated by machine, the stone dust is impacted into every nook and cranny forming a fire resistant coating as well as neutralising any coal dust present.

94   In the course of inspections during the year, HM Inspectors collected 349 roadway dust samples, of which 30 or 8·6% were below the prescribed standard. Immediate action was taken to have these places treated and the matter was taken up with the management.

Training and education

95   As referred to earlier in the report, a successful recruitment drive brought an increased number of men into the industry during the year. Preliminary training courses were completed by 2975 men attending the training centres, one of which had to be extended to accommodate the increased numbers. Craft apprenticeship courses were completed by 13 mining, 58 mechanical engineering and 68 electrical engineering students. Coal face training and improvership training was completed by 618 men.

96   To meet particular needs, many other training courses were arranged for groups of men such as potential deputies; coal face trainee supervisors; underground loco drivers, shunters and guards; crane operators and slingers; and air borne dust samplers. In addition, refresher courses were held for officials, mechanics and electricians and instructional courses for men employed on coal faces where nucleonic steering devices are used.

97   Accident investigations have brought to light a serious deficiency in the current training programme which will have to be revised. This has arisen by sheer weight of numbers and circumstances not forseen when the existing training schemes were devised. The statistics show that at the end of the year out of a total of 29 200 men employed in coal mines in the district, more than 4000 had less than 2 years experience of underground working, and a further 1500 were men who had re-entered the industry within that period. Most of these 5500 men were employed underground, and by the ordinary course of events, nearly all would be employed on haulage and transport work. This concentration of inexperienced men is inevitable as haulage and transport work has long been the traditional training ground for miners where they could work alongside experienced men and gain general pit sense.

98   An added complication is the NCB early retirement scheme. Last year alone, 1181 long service miners took advantage of the scheme to retire at 63 years of age instead of 65, and it is the intention to progressively reduce the age limit to 60 years. This effectively takes out of circulation many man-years of valuable experience, and again in the natural course of events most of these men would have been employed on haulage and transport duties.

99   The net result is a much higher proportion of inexperienced men on haulage and transport work, and in a small team of men operating a particular supply system there may be no-one sufficiently competent to take charge and supervise the work. The current training schemes were not designed to accommodate this situation, training periods were not specified and the qualifications of the supervisor were not adequately defined. It appears there is now the need to devise a haulage and transport training programme similar to that scheme already in operation for coal face training, with a similar 'improvership' training at the man's place of work and additional training when a man is transferred to a different type of haulage system.

100   No new legislation is necessary. Section 88 of the Mines and Quarries Act 1954 already requires the manager to ensure all persons he employs is adequately trained in the task being undertaken.

Health and welfare

Noise

101   There has been little progress in the suppression of noise at source but the principles of enclosure by sound absorbent materials or the removal of the operator to a sound proof enclosure have been applied where possible. Also the noise factor is now taken into account when new machinery is installed. The NCB have noise meters and trained personnel together with medical advisors to investigate and advise on particular problems during operational and planning stages. This awareness should be furthered by demanding from manufacturers a supply of products which produce a minimum of noise at a tolerable sound frequency.

Medical services

102   The high standard of medical service was maintained throughout the year, minimising the pain and suffering of anyone unfortunate enough to be injured or be taken ill on the mine premises. The increased recruitment caused a heavy workload of medical examinations by the doctors and their staff but they still found time to encourage and advise in first aid training and competitions. There was no shortage of first aid volunteers for training or improving their expertise, and the first aid competitions were well attended.

103   The welfare arrangements at each mine are good with pit head baths and canteens at all mines. In addition there are facilities at most mines for specialised clubs if a sufficient number of employees show interest. In this respect, brass bands are popular, and Desford Colliery Band won the NCB trophy at the National Contest in Blackpool in November. To promote safety, the NCB organised Safety Sketch Competitions and the final competition was won by Bagworth Mine employees. A third and neighbouring mine, Ellistown, nearly made it a triple win when they came second in the national final of the Surface Safety and Tidiness Competition.

104   I congratulate not only the winners of the competitions but all who took part as competitors, organisers or judges. Anything which promotes safe working and maintains interest in safety must be good for the industry.

3   Mines of stratified ironstone, shale and fireclay

105   The last of the fireclay mines in the district was abandoned during the early part of the year having finally succumbed to the economic pressure. No replacement mine is planned, the remaining clay will be extracted by opencast quarrying.

4   Miscellaneous mines

106   The two mines in this category continued working throughout the year producing a high quality gypsum for use in their nearby factories. Output has been reasonably consistent although the demand for plaster, plaster board and allied products was low in common with other building materials.

There was no change in the method of mining nor in the working conditions. Extensive reserves of mineral have been proved, and a third mine is being planned as an eventual replacement of the older of the two working mines.

Accidents and dangerous occurrences

107   There was no accident nor any dangerous occurrence reported during the year compared with two persons seriously injured and one outbreak of fire last year. When figures are so low trends do not make sense but nevertheless for a

company employing men at two mines and three quarries in the district to have a year free from serious accidents and dangerous occurrences, the achievement is well worth placing on record.

Health and welfare

108   The investigation into the origin and dispersal of radon and radon daughters was continued during the year and although seemingly little real progress has been made the owners are to be commended in sustaining their efforts at one mine where the measured levels were found to be significant to minimise the exposure of workers to this hazard. Improved coursing of the circulating air in the mine and planned changes in the ventilation system should alleviate the situation without, it is hoped, resorting to filtration.

109   As was mentioned last year a survey was made using several methods of sampling. The results were not too consistent when comparing instruments but the ease and speed of a portable instant read off meter made it very acceptable and useful for comparing results in a before and after situation when corrective action is deemed necessary.

110   All the employees are subject to an annual chest X-ray examination and their general health monitored, but nothing has been observed to date which would suggest anything untoward.

111   At both mines suitable arrangements are made for changing and drying clothes and bathing. First aid facilities are more than adequate, and men are encouraged to take an interest in first aid training.

Fire prevention

112 Efforts have continued with some success towards the eventual elimination of the major fire hazards associated with the vehicles in use underground, namely, use of mineral oil in hydraulic circuits and heavy duty batteries for starting the diesel engines. Rubber conveyor belting has not been used underground for a number of years.

113   Rescue services are organised in conjunction with the local NCB central mines rescue station where the gypsum miner's team is trained.

5 Quarries

General

114 There was little change in the total number of quarries working at the end of the year compared with 1976 as shown in the table below:-

Table e. Number of quarries worked 1976/77

| | Number of quarries | |
Mineral	1976	1977
Coal	13	11
Fireclay	48	47
Ironstone	12	12
Chalk and limestone	44	42
Igneous rock	17	17
Sand and gravel (including Silica sand)	94	93
Sandstone	11	11
Gypsum	6	6
Miscellaneous (including Slag shale)	4	7
Total	**249**	**246**

115 Quarrying generally was affected by the depressed state of other industries though there were more optimistic signs towards the end of the year. Manpower levels remained fairly static although there was some short-time working at quarries where excessive stocks had accumulated. Only six of the opencast coal sites were in regular production, the remaining registered sites being temporarily stopped for development or awaiting planning permission and licence to work the remaining mineral. Quarries producing fireclay, gypsum and block limestone were kept reasonably busy during the year, but output of other minerals was below average. Again this year, economic considerations held back developments of new projects and modernisation at most quarries although a new plant was installed at a recently opened sand and gravel quarry.

Environment

116 Ground conveyors were installed in two quarries, replacing dump trucks in one and a standard gauge railway system in the other. These changes however, were strongly resented by the nearby residents culminating in complaints of introduced hazards to children attempting to steal rides on the moving conveyors and of changes in noise level, particularly the sounding of pre-start warning devices in one case. Tests revealed that the latter irritation was most probably due to sound frequency rather than an unacceptable sound level. Noise levels, potential hazards to children and inadequacy of fencing on

the premises had not been subjects of complaint until the changes of the system of transport were made.

117   Only one blasting complaint was received and this was from residents in a very localised area some 4km away from the horseshoe shaped quarry face although at and near the quarry noise levels were normal. The explanation would appear to be that the sound waves encountered a thermal inversion, warm air above cooler air, and unattenuated were focussed downwards in a very narrow band over intervening high ground and producing concentrated sound levels back at ground level over a very restricted area.

118   Due to 'scatter' experienced in the present range of delay detonators readily available to quarries it is inevitable that irregularities in the time delay reflected in more concentrated peak shock waves will draw adverse comment and possible complaint from residents in the affected zone. At one quarry constantly faced with this problem trials were carried out using an electronic sequential timing device to initiate each hole at exactly 50 millisecond intervals. The uniformity of the timed delay resulted in approximately 50% reduction in vibration as recorded on an instrument within a nearby private property. The experiment was repeated with a further round of shots with similar results. Extended trials are to be continued during next year and should furnish sufficient evidence and records to draw firmer conclusions.

119   The management of the quarry concerned have extensive comprehensive documentation of all shotfiring for a considerable time past during which their practice and techniques have evolved to counter their particular environment and at this point in time the adoption of sequential shotfiring will be considered only at quarries subjected to similar environmental problems.

120   Visible dust being raised during quarrying operations is the prevalent source of the public complaints and in one instance the local authority commissioned a survey for airborne dust to be carried out around the perimeter of a quarry under surveillance. Airborne dust surveys were carried out within the curtilage by Inspectorate staff and the quarry owners also to pin point high makes of dust.

121   Check samples give an indication only of the conditions prevailing at the time and can vary widely with changes in operation and climatic conditions and particularly with regard to wind speed and direction where nearby dwellings may be involved. Frequent regular monitoring, especially in hard rock quarries, is the only means of determining that employees are not exposed to unacceptable levels of respirable dust, if this is achieved within the curtilage then no complaints should arise from without. This must be the aim of every management but regretably not all have provided themselves with the facilities and trained personnel to do so and are failing to comply both with the spirit

and wording of Sections 2 and 3 of the Health and Safety at Work Etc Act 1974. Pollution is an aspect, be it noise or dust, which will be given serious and increasing consideration by Local Authorities when applications are made for planning permission for quarry extensions or the opening of new quarries.

122   Managements are requested to study carefully and then implement the recommendations put forward in the HSE publication *'*Airborne Dust in Quarries - Health Precautions*' if they are intent, as they must be, to safeguard the health of employees and non-employees in respect of respirable dust.

Accidents and their prevention

123   Four persons were seriously injured compared with eight in 1976. Two of the accidents occurred in one opencast coal quarry.

124   At a large granite quarry, a shunter was crushed between the buffers of two 27 tonnes capacity railway wagons full of stone. There was no witness to the accident but it appears the injured man went between a stationary and a moving wagon in order to free a flexible brake pipe fouling the coupling chain, and he mis-judged the speed of the oncoming vehicle.

125   At an opencast coal and fireclay quarry, a driver, assisting a mechanic working on a large motorised scraper suddenly collapsed and complained of partial paralysis in his limbs. Apart from a slight graze on his head, there was no obvious injury and it appears the driver simply struck his head on part of the vehicle. Wearing a safety helmet probably would have prevented his injury.

126   At the same opencast coal and fireclay quarry, a maintenance fitter was carrying out electrical repairs at the front end of a large earth mover vehicle when he was struck and seriously injured by the rear back-acter bucket of a passing tractor shovel. The vehicle being repaired had been parked on a recognised running road instead of being taken round to the workshop, but in any case, the parked vehicle should have been better illuminated to avoid collisions.

127   The fourth accident occurred at a clay quarry when a labourer stumbled and fractured his leg, his boot lace having caught in a small tree root lying across his path.

Dangerous occurrences

128   One dangerous occurrence was reported during the year compared with two in 1976. At an igneous rock quarry, after a period of heavy rains, the

---

*Available from HMSO price 50p ISBN 011880500 2.

outside bank of one of a series of lagoons became unstable and subsided to a depth of 1m over a length of 30m forming a shallow circular slip type of failure. No damage or injury was caused, and the bank was simply strengthened over the full length of the failure zone.

Engineering

129   In general, standards of engineering are improving at the larger quarries where installed expensive machinery is maintained to an adequate planned maintenance scheme. But in some quarries there is a tendency to delay repairs to internal user vehicles for economic reasons only, perhaps to the extent of infringing on safety before deciding to scrap the vehicle. This policy cannot be condoned in the interest of safety of the employees.

130   During the year, two incidents which caused concern were not reportable under Section 116 of the Mines and Quarries Act 1954 simply because by good fortune no one was injured. In both cases there was failure of welded bridging structures across a public road.

131   After being subjected to gale force winds an overhead conveyor gantry completely collapsed and fell partly on the quarry premises and partly on a public road. The investigation revealed possible errors in the basic design and suspect welding technique during manufacture. It also appeared modifications had been carried out without reference to the original design.

132   The second structural failure was at a quarry working at each side of a public road spanned by a Bailey bridge of war time fame but of recent manufacture. Fatigue failure cracks were discovered in some of the 26 welded steel transoms, and after tests had been carried out, it was decided to replace all of them.

133   By strict regulation only one dumper, 70 tonnes gross tare passed over the bridge at any one time, albeit at intervals of 10 seconds, keeping the actual loading within the designed capacity of 100 tonnes. Estimations of the number of load cycles however, suggested that the bridge structure may have been subjected to more than that to which it was designed and resulting in the observed metal fatigue. Rigid control of vehicular movements, recording of load cycles imposed and a thorough investigation into the design of the transoms, their location and method of attachment should avoid further incidents.

Civil engineering

134   As already noted earlier in the report there was only one notification of instability arising at a lagoon or tip.

135   Tips and lagoons are now accepted as civil engineering structures, outside consultants being engaged at many quarries to design, supervise the construction and make regular reports thereon as is required by legislation. The criticisms made during inspections arise mainly from the day to day operations and stem from lack of basic knowledge at local management level. All too frequently attention has to be drawn to excessive localised tipping, lack of consolidation and inadequate drainage of surface water from tips. Similarly with lagoons, reductions of freeboard arising from blocked or wrongly sited overflow pipes causing wave erosion of the lagoon retaining walls is a common fault. In some places retaining walls have been found reduced to half their original width by this erosion.

136   Managers are required to give close and effective supervision to all operations within their quarry including those at tips and lagoons. The observed failure to keep records and reports up to date, to adhere rigidly to tipping rules and apparent lack of observation of vital matters which could affect the stability of a lagoon or tip, suggests that quarry owners and senior management should carry out a re-appraisal of their instructions concerning tips and lagoons, and ensure all the quarry managers are trained and competent to carry out the instructions.

137   A further regular criticism is that in larger concerns, senior management staff draw up a generalised form of Tipping rules but do not always ensure that the rules cater for each individual site. Rules which cannot be applied defeat the whole object of this legislation and discredit those which are applicable.

Health and welfare

138   Airborne dust remains a major health hazard in quarries and for this reason inspectors continue to take check samples at selected spots, especially where the dust is likely to have a high quartz content, to highlight the danger areas and evaluate any progress made by management in response to their statutory duty to minimise the make of harmful dust. The issue of respirators to some workers who may be exposed to high concentrations of dust, and the often heard claim that in normal circumstances workmen need not enter heavy dust laden atmospheres found in preparation plants, are not the solutions nor the answers to the legal requirement to minimise the make of dust. More can and needs to be done towards the reduction of the airborne dust at the source, improved control of dust at points where it may become airborne, regular monitoring at and around these points and more frequent cleaning up to remove dust that may be disturbed underfoot.

139   A second quarry is now equipped for steam dust suppression at a conveyor delivery discharge point. The efficacy of this measure is still being assessed.

140   Personal dust sampling equipment has been used at several large quarries by HM Inspectors of Quarries and the quarries management staff in an attempt to evaluate the problem. Much can be done to minimise the make of dust in a quarry particularly at crushers, screening plant and drilling rigs. If dust is trapped where it is being made it is easier to control than when it becomes air borne.

141   At three quarries trials were carried out with safety helmets fitted with clear visors and battery powered fans to pass a flow of filtered air across the wearer's face. These may have limited use where other methods of dust control are impracticable, but they will not be issued generally to all workers.

142   Extensive noise level surveys have been carried out in some quarries and where necessary attempts were made to suppress the noise or to remove the workman from the noisy environment. When new machinery is installed, noise levels should be taken into account from the outset as generally it is easier to avoid a problem than to cure it.

143   At most quarries workmen are now issued with safety helmets together with other safety equipment and protective clothing but there seems to be some reluctance to wearing the equipment provided. Management staff and Trade Union officials must set an example here and take up the point with anyone they find not properly attired in the quarry.

144   At least the minimum first-aid equipment is available at all quarries, and logically the larger the quarry, the better the facilities. Mess rooms, toilet and drying room accommodation similarly varies according to the size of the undertaking, but in some cases, a little care and thought exercised by the men using the facilities could quickly improve standards of hygiene.

145   Training courses and safety meetings were again arranged by the quarry owners and managers, plant manufacturers and quarry organisations such as the Institute of Quarrying and the Ceramics Glass and Mineral Products Industry Training Board. In addition a quarryman's shotfiring course was held at a local technical college. All were well attended and were considered a useful contribution to safe working but they have only a local effect.

146   Although not required by regulations a more general training scheme is necessary for the whole industry. Section 88 of the Mines and Quarries Act 1954 requires that no person should be employed in a quarry unless he has received adequate instruction and (where necessary) training for his work. The interpretation of this requirement varies and in some places the training is

negligible. A minimum period of working under supervision, say 20 working days, should be stipulated by the management and applied not only to new employees but also to men transferred to other types of work within the quarry. Records should also be kept as evidence of that training.

6 New entrants

147 'New entrants' is a term applied to certain working places not previously governed by direct legislation but now included in the general application of the Health and Safety at Work etc Act 1974 and attached to the appropriate Inspectorate. In particular the new entrants we are currently concerned with are all mineral exploratory drilling sites and landfill sites on mine or quarry premises.

Landfill sites

148 This term also requires explanation. It was introduced to distinguish between tips of waste material from mines and quarries already covered by adequate legislation, and other refuse tips where extraneous waste material is to be put down on mine or quarry premises. HM Inspectors of Mines and Quarries make inspections of the sites on being notified by the licensing authority that they propose to issue a licence subject to certain conditions. Landfill sites are controlled by licence under the Control of Pollution Act 1974 issued by the Local Authority who are mainly concerned with the pollution aspects of the proposed tip. The inspection of the site is then concerned with the health and safety aspects in relation to the persons employed at the tip, the safety of the operational procedures and the safety and health of persons employed in any working mines or quarries in the vicinity. Following such an inspection, the terms of the licence may be modified to avoid any foreseen hazard.

149 During the year 90 notifications of licence applications were dealt with and the sites inspected but it is anticipated the number will rapidly increase once the system of notification has been streamlined and the routine established.

Exploratory drilling

150 Exploratory drilling is a normal mining operation and where this is done on mine or quarry premises, or underground, the Mines and Quarries Acts and Regulations apply but recently there has been an upsurge in exploration

activity in the search for coal and other minerals well beyond the existing mining areas. Forty four notifications of commencement of exploratory drilling operations were received during the year and significant numbers of men are now employed in this work drilling deep holes over a wide geographical area. As the holes are drilled deeper the drilling rigs and ancillary equipment has to be made bigger, heavier and more powerful, introducing hazards into what was hitherto a simple, small scale operation.

151   Typical of the hazards magnified by the scale of current operations was a fatal accident which occurred at one drilling site. A large rig had been set up and drilling had commenced when a double load of tubular steel drill rods was delivered by articulated lorry and trailer. To take the empty trailer back it was being loaded on to the articulated lorry by means of a mobile crane but in the process the front wheel bogey of the trailer became detached, fell on to a man and killed him. Perhaps the accident could not be directly attributed to the drilling operation, but it brought out the fact that the transporting and handling of this heavy, bulky equipment needs careful organisation and management.

152   A further complication is that work on this scale attracts contractors, sub-contractors with supporting service contractors and there may be some doubt as to who is in charge. Although there is no legal requirement to appoint a manager, the main contractor should appoint a person to take charge and co-ordinate all the work on the site.

153   In the course of inspection of the sites standards of operation were found to be generally satisfactory with adequate health and welfare facilities provided and a good approach towards safe working.

154   One of the many drill rig operating firms makes awards to employees for safe working and have 'on site' safety meetings every week. The matters discussed, any suggestions made and action taken are recorded in the minutes of the safety meeting, and this is supplemented by a quarterly detailed inspection by the Field Supervisor who completes a comprehensive check list, again for the record. The approach to safety incorporating a two-way discussion between employer and employee, thus meeting the requirements of Section 2 of the Health and Safety at Work etc Act 1974, is recommended for adoption by other operators. Safety is everyone's business.

Appendix 1 Staff in post at 31 December 1977

HM Senior District Inspector of Mines and Quarries	F Tootle
HM District Inspectors of Mines and Quarries	A G Conliffe P Williams *G Scott
HM Inspectors of Mines and Quarries	R A K Colquhoun K L Twist
HM Electrical Inspector of Mines and Quarries	W Laing
HM Inspector of Mechanical Engineering	H W Morrell
HM Inspector of Quarries	R J Gerrett †G Henderson

*Headquarters outstationed staff
†Also inspects quarries in North Midlands District.

Appendix 2 Mines and Quarries:

Number of fatal and serious (reportable) accidents by type and location 1974-77	Number of fatal accidents		Number of serious (reportable) accidents	
Classification of Accident	*Average 1974-76*	*1977*	*Average 1974-76*	*1977*
COAL MINES				
Falls of ground:				
(a) At the face	1	1	9	16
(b) On the roads	—	—	1	—
Haulage and transport	1	1	19	27
Machinery	1	1	3	6
Explosives	—	—	1	—
Gases, coal dust and fires				
(a) Explosions	—	—	—	—
(b) Others	—	—	—	—
Other underground accidents	1	—	14	10
Shafts	—	—	⨍	—
Surface accidents	—	1	7	4
Total coal mines	4	4	53	63
MINES OF STRATIFIED IRONSTONE, SHALE AND FIRECLAY	—	—	—	—
MISCELLANEOUS MINES	—	—	1	—
QUARRIES				
(a) Opencast coal	—	—	—	2
(b) Others	1	—	5	2
Total all mines and quarries	5	4	59	67

Note: In the tables averages have been rounded off and any figure less than 0.5 represented by the symbol ⨍.

Source: Health and Safety Executive.

Appendix 3   Coal Mines

Number of fatal and serious (reportable) accidents by type and location, 1974-77	Number of fatal accidents		Number of serious (reportable) accidents	
	Average 1974-76	*1977*	*Average 1974-76*	*1977*
UNDERGROUND ACCIDENTS				
FALLS OF GROUND				
Face				
1(a) Roof	†	1	4	9
1(b) Face or side	†	—	5	7
Total	1	1	9	16
Elsewhere underground				
1(a) Roof	—	—	†	—
1(b) Face or side	—	—	1	—
Total	—	—	1	—
Total falls of ground	1	1	10	16
TRANSPORT				
Face				
2 Rope haulage	—	—	1	1
6 Chain conveyors	†	—	1	4
3, 4, 5 and 7, 8, 9 others	—	—	†	—
Total	†	—	3	5
Elsewhere underground				
2 Rope haulage				
2(a) Breakages	—	—	1	—
2(b) Runaways	—	—	1	1
2(c) Hold-fasts	—	—	1	2
2(d) De-railment or re-railing	—	—	1	3
2(e) Struck by vehicle other than (a) to (d)	—	—	1	3
2(f) Ropes and pulleys	—	—	1	5
2(g) Others	†	—	1	3
3 Hand or gravity haulage	—	—	1	—
4 Locomotive haulage	—	—	2	2
6 Chain conveyors	—	1	†	1
7 Belt conveyors	—	—	4	2
5, 8, 9 Others	—	—	†	—
Total	†	1	16	22
Total transport	1	1	19	27

Appendix 3 (continued)

Accident classification	Number of fatal accidents		Number of serious (reportable) accidents	
	Average 1974-76	*1977*	*Average 1974-76*	*1977*
MACHINES				
Face				
10 Power loaders	1	—	1	4
11 Coal cutters	—	—	—	—
12, 13 Roadway cutter loaders, ripping machines, mechanical loaders	≁	—	1	2
14 Machines other than 10 to 13	—	—	≁	—
Total	1	—	2	6
Elsewhere underground				
12, 13 Roadway cutter loaders, ripping machines, mechanical loaders	≁	—	1	—
14 Machines other than 12 and 13	—	1	—	—
Total	≁	1	1	—
Total machines	2	1	2	6
MISCELLANEOUS				
15 Powered supports	—	—	1	1
16 Use of tools or appliances (including flying splinters)	—	—	1	2
17 Falling objects (excluding falls of ground)	≁	—	3	3
18 Stumbling, falling or slipping				
18 (a) slipping or falling	—	—	1	—
18 (b) falling over fixed equipment	—	—	2	—
18 (c) falling over other equipment or material	—	—	1	—
18 (d) falling from a height greater than 1 metre	—	—	2	2
18 (e) Others	≁	—	1	—
19 Explosives	—	—	1	—
20 Electricity	—	—	—	—
21 Inrushes of water	—	—	—	—
22 Explosions of firedamp or coal dust	—	—	—	—
23 Suffocation	—	—	—	—
24 Fires	—	—	—	—
27 All other underground accidents				
(A) Face	≁	—	1	1
(B) Elsewhere	—	—	1	1
Total miscellaneous	1	—	15	10
ALL SHAFT AND STAPLE PIT ACCIDENTS	—	—	≁	—
Total underground accidents	4	3	47	59

Accident classification	Number of fatal accidents		Number of serious (reportable) accidents	
	Average 1974-76	*1977*	*Average 1974-76*	*1977*
SURFACE ACCIDENTS				
TRANSPORT				
3 Hand or gravity haulage	—	—	†	—
5 Trackless haulage	—	—	†	—
6-8 Conveyors	—	1	†	—
2, 4, 9 Others	—	—	1	—
MACHINES				
10-14 Machines	—	—	1	2
MISCELLANEOUS				
Appliances (including flying splinters)	—	—	—	—
17 Falling objects	—	—	1	2
18 Stumbling, falling or slipping	—	—	1	—
20 Electricity	—	—	—	—
15, 19, 22, 23, 24, 26 and 27 All others	—	—	2	—
Total surface accidents	—	1	7	4
Total Accidents	4	4	53	63

Source: Health and Safety Executive

Printed in England for Her Majesty's Stationery Office by Hobbs the Printers of Southampton
(1690) Dd022509 K8 10/78 G3927